Company Name: _____

Address: _____

City, State, Zip _____

Contact Name: _____

Phone Number: _____

Email: _____

Log Dates / Year: _____

DATE:		DAY:	S	M	T	W	T	F	S

Foreman:

Project Details

Weather	
High	Low

Weather Conditions

Safety Meeting		
Yes	No	Subject:

Concerns:

Ground Conditions

Visitors

Schedule	
Completion Date	
Days Ahead	
Days Behind	

Accidents / Injuries / Delays

Summary of Work Done Today

Signature: Title:

Employee	Trade	Hours	OT	Subcontractors	Trade	OT

Equipment	#	On	Off

Material Delivered	#	Rentals	Rented From/Rate

DATE:		DAY:	S M T W T F S

Foreman:

Project Details		Weather	
		High	Low
		Weather Conditions	

Safety Meeting

Yes	No	Subject:

Concerns:

	Ground Conditions

Visitors

Schedule

Completion Date	
Days Ahead	
Days Behind	

Accidents / Injuries / Delays

Summary of Work Done Today

Signature: Title:

Employee	Trade	Hours	OT	Subcontractors	Trade	OT

Equipment	#	On	Off

Material Delivered	#	Rentals	Rented From/Rate

DATE:	DAY:	S M T W T F S

Foreman:

Project Details	Weather

	High	Low

Weather Conditions

Safety Meeting

Yes	No	Subject:

Concerns:

Ground Conditions

Visitors	Schedule

Completion Date	
Days Ahead	
Days Behind	

Accidents / Injuries / Delays

Summary of Work Done Today

Signature:	Title:

Employee	Trade	Hours	OT	Subcontractors	Trade	OT

Equipment	#	On	Off

Material Delivered	#	Rentals	Rented From/Rate

DATE:		DAY:	S M T W T F S

Foreman:

Project Details	Weather	
	High	Low

Weather Conditions

Safety Meeting

Yes	No	Subject:
Concerns:		

Ground Conditions

Visitors	Schedule	
	Completion Date	
	Days Ahead	
	Days Behind	

Accidents / Injuries / Delays

Summary of Work Done Today

Signature: Title:

Employee	Trade	Hours	OT	Subcontractors	Trade	OT

Equipment	#	On	Off

Material Delivered	#	Rentals	Rented From/Rate

DATE: **DAY:**   S   M   T   W   T   F   S

Foreman:

Project Details

Weather

High	Low

Weather Conditions

Ground Conditions

Safety Meeting

Yes	No	Subject:

Concerns:

Visitors

Schedule

Completion Date	
Days Ahead	
Days Behind	

Accidents / Injuries / Delays

Summary of Work Done Today

Signature: **Title:**

Employee	Trade	Hours	OT	Subcontractors	Trade	OT

Equipment	#	On	Off

Material Delivered	#	Rentals	Rented From/Rate

DATE:	DAY:	S M T W T F S

Foreman:

Project Details

Weather

High	Low

Weather Conditions

Ground Conditions

Safety Meeting

Yes	No	Subject:

Concerns:

Visitors

Schedule

Completion Date	
Days Ahead	
Days Behind	

Accidents / Injuries / Delays

Summary of Work Done Today

Signature:	Title:

Employee	Trade	Hours	OT	Subcontractors	Trade	OT

Equipment	#	On	Off

Material Delivered	#	Rentals	Rented From/Rate

DATE:		DAY:	S	M	T	W	T	F	S

Foreman:

Project Details	Weather	
	High	Low

	Weather Conditions

Safety Meeting

Yes	No	Subject:

Concerns:

	Ground Conditions

Visitors

Schedule

Completion Date	
Days Ahead	
Days Behind	

Accidents / Injuries / Delays

Summary of Work Done Today

Signature: Title:

Employee	Trade	Hours	OT	Subcontractors	Trade	OT

Equipment	#	On	Off

Material Delivered	#	Rentals	Rented From/Rate

DATE: **DAY:**   S   M   T   W   T   F   S

Foreman:

Project Details	Weather	
	High	**Low**

Weather Conditions

Safety Meeting

Yes	No	Subject:

Concerns:

Ground Conditions

Visitors

Schedule

Completion Date	
Days Ahead	
Days Behind	

Accidents / Injuries / Delays

Summary of Work Done Today

Signature: **Title:**

Employee	Trade	Hours	OT	Subcontractors	Trade	OT

Equipment	#	On	Off

Material Delivered	#	Rentals	Rented From/Rate

| DATE: | | DAY: | S | M | T | W | T | F | S |

Foreman:

Project Details

Weather	
High	Low

Weather Conditions

Ground Conditions

Safety Meeting

Yes	No	Subject:

Concerns:

Visitors

Schedule	
Completion Date	
Days Ahead	
Days Behind	

Accidents / Injuries / Delays

Summary of Work Done Today

Signature: Title:

Employee	Trade	Hours	OT	Subcontractors	Trade	OT

Equipment	#	On	Off

Material Delivered	#	Rentals	Rented From/Rate

DATE:		DAY:	S M T W T F S

Foreman:

Project Details

Weather

High	Low

Weather Conditions

Ground Conditions

Safety Meeting

Yes	No	Subject:

Concerns:

Visitors

Schedule

Completion Date	
Days Ahead	
Days Behind	

Accidents / Injuries / Delays

Summary of Work Done Today

Signature: _____ Title: _____

Employee	Trade	Hours	OT	Subcontractors	Trade	OT

Equipment	#	On	Off

Material Delivered	#	Rentals	Rented From/Rate

DATE:		DAY:	S	M	T	W	T	F	S

Foreman:

Project Details	Weather	
	High	Low

	Weather Conditions

Safety Meeting

Yes	No	Subject:	Ground Conditions
Concerns:			

Visitors

	Schedule	
	Completion Date	
	Days Ahead	
	Days Behind	

Accidents / Injuries / Delays

Summary of Work Done Today

Signature: Title:

Employee	Trade	Hours	OT	Subcontractors	Trade	OT

Equipment	#	On	Off

Material Delivered	#	Rentals	Rented From/Rate

DATE:		DAY:	S	M	T	W	T	F	S

Foreman:

Project Details	Weather	
	High	Low
	Weather Conditions	

Safety Meeting

Yes	No	Subject:
Concerns:		

Ground Conditions

Visitors

	Schedule	
	Completion Date	
	Days Ahead	
	Days Behind	

Accidents / Injuries / Delays

Summary of Work Done Today

Signature: Title:

Employee	Trade	Hours	OT	Subcontractors	Trade	OT

Equipment	#	On	Off

Material Delivered	#	Rentals	Rented From/Rate

DATE:	DAY:	S	M	T	W	T	F	S

Foreman:

Project Details

Weather

High	Low

Weather Conditions

Ground Conditions

Safety Meeting

Yes	No	Subject:

Concerns:

Visitors

Schedule

Completion Date	
Days Ahead	
Days Behind	

Accidents / Injuries / Delays

Summary of Work Done Today

Signature: _____ Title: _____

Employee	Trade	Hours	OT	Subcontractors	Trade	OT

Equipment	#	On	Off

Material Delivered	#	Rentals	Rented From/Rate

DATE:		DAY:	S	M	T	W	T	F	S

Foreman:

Project Details

Weather	
High	Low

Weather Conditions

Safety Meeting		
Yes	No	Subject:

Concerns:

Ground Conditions

Visitors

Schedule	
Completion Date	
Days Ahead	
Days Behind	

Accidents / Injuries / Delays

Summary of Work Done Today

Signature:	Title:

Employee	Trade	Hours	OT	Subcontractors	Trade	OT

Equipment	#	On	Off

Material Delivered	#	Rentals	Rented From/Rate

DATE:		DAY:	S	M	T	W	T	F	S

Foreman:

Project Details	Weather

	High	Low

	Weather Conditions

Safety Meeting

Yes	No	Subject:

Concerns:

	Ground Conditions

Visitors

	Schedule

Completion Date	
Days Ahead	
Days Behind	

Accidents / Injuries / Delays

Summary of Work Done Today

Signature: _____ Title: _____

Employee	Trade	Hours	OT	Subcontractors	Trade	OT

Equipment	#	On	Off

Material Delivered	#	Rentals	Rented From/Rate

DATE:		DAY:	S M T W T F S

Foreman:

Project Details	Weather	
	High	Low

	Weather Conditions

Safety Meeting

Yes	No	Subject:
		Concerns:

Ground Conditions

Visitors	Schedule	
	Completion Date	
	Days Ahead	
	Days Behind	

Accidents / Injuries / Delays

Summary of Work Done Today

Signature: Title:

Employee	Trade	Hours	OT	Subcontractors	Trade	OT

Equipment	#	On	Off

Material Delivered	#	Rentals	Rented From/Rate

| DATE: | | DAY: | S | M | T | W | T | F | S |

Foreman:

Project Details		Weather	
		High	Low

	Weather Conditions

Safety Meeting

Yes	No	Subject:

Concerns:

	Ground Conditions

Visitors

Schedule

Completion Date	
Days Ahead	
Days Behind	

Accidents / Injuries / Delays

Summary of Work Done Today

Signature: Title:

Employee	Trade	Hours	OT	Subcontractors	Trade	OT

Equipment	#	On	Off

Material Delivered	#	Rentals	Rented From/Rate

DATE:		DAY:	S	M	T	W	T	F	S

Foreman:

Project Details	Weather	
	High	**Low**

Safety Meeting

Yes	No	Subject:

Concerns:

Weather Conditions

Ground Conditions

Visitors

Schedule

Completion Date	
Days Ahead	
Days Behind	

Accidents / Injuries / Delays

Summary of Work Done Today

Signature: Title:

Employee	Trade	Hours	OT	Subcontractors	Trade	OT

Equipment	#	On	Off

Material Delivered	#	Rentals	Rented From/Rate

DATE:		DAY:	S	M	T	W	T	F	S

Foreman:

Project Details

Weather

High	Low

Weather Conditions

Ground Conditions

Safety Meeting

Yes	No	Subject:

Concerns:

Visitors

Schedule

Completion Date	
Days Ahead	
Days Behind	

Accidents / Injuries / Delays

Summary of Work Done Today

Signature: Title:

Employee	Trade	Hours	OT	Subcontractors	Trade	OT

Equipment	#	On	Off

Material Delivered	#	Rentals	Rented From/Rate

DATE:		DAY:	S	M	T	W	T	F	S

Foreman:

Project Details

Weather

High	Low

Weather Conditions

Ground Conditions

Safety Meeting

Yes	No	Subject:

Concerns:

Visitors

Schedule

Completion Date	
Days Ahead	
Days Behind	

Accidents / Injuries / Delays

Summary of Work Done Today

Signature: _____   Title: _____

Employee	Trade	Hours	OT	Subcontractors	Trade	OT

Equipment	#	On	Off

Material Delivered	#	Rentals	Rented From/Rate

| DATE: | | DAY: | S | M | T | W | T | F | S |

Foreman:

Project Details

Weather

High	Low

Weather Conditions

Ground Conditions

Safety Meeting

Yes	No	Subject:

Concerns:

Visitors

Schedule

Completion Date	
Days Ahead	
Days Behind	

Accidents / Injuries / Delays

Summary of Work Done Today

Signature: Title:

Employee	Trade	Hours	OT	Subcontractors	Trade	OT

Equipment	#	On	Off

Material Delivered	#	Rentals	Rented From/Rate

DATE:		DAY:	S	M	T	W	T	F	S

Foreman:

Project Details	Weather	
	High	Low

	Weather Conditions
Safety Meeting	

Yes	No	Subject:

Concerns:

Ground Conditions

Visitors	Schedule	
	Completion Date	
	Days Ahead	
	Days Behind	

Accidents / Injuries / Delays

Summary of Work Done Today

Signature:	Title:

Employee	Trade	Hours	OT	Subcontractors	Trade	OT

Equipment	#	On	Off

Material Delivered	#	Rentals	Rented From/Rate

DATE:		DAY:	S M T W T F S

Foreman:

Project Details

Safety Meeting

Yes	No	Subject:

Concerns:

Visitors

Weather

High	Low

Weather Conditions

Ground Conditions

Schedule

Completion Date	
Days Ahead	
Days Behind	

Accidents / Injuries / Delays

Summary of Work Done Today

Signature:	Title:

Employee	Trade	Hours	OT	Subcontractors	Trade	OT

Equipment	#	On	Off

Material Delivered	#	Rentals	Rented From/Rate

DATE:	DAY:	S M T W T F S

Foreman:

Project Details	Weather	
	High	Low

Weather Conditions

Safety Meeting

Yes	No	Subject:

Concerns:

Ground Conditions

Visitors

Schedule

Completion Date	
Days Ahead	
Days Behind	

Accidents / Injuries / Delays

Summary of Work Done Today

Signature:	Title:

Employee	Trade	Hours	OT	Subcontractors	Trade	OT

Equipment	#	On	Off

Material Delivered	#	Rentals	Rented From/Rate

DATE:		DAY:	S M T W T F S

Foreman:

Project Details		Weather	
		High	**Low**

Weather Conditions

Safety Meeting				
Yes	No	Subject:		
Concerns:				

Ground Conditions

Visitors

Schedule	
Completion Date	
Days Ahead	
Days Behind	

Accidents / Injuries / Delays

Summary of Work Done Today

Signature:	Title:

Employee	Trade	Hours	OT	Subcontractors	Trade	OT

Equipment	#	On	Off

Material Delivered	#	Rentals	Rented From/Rate

DATE:

DAY: S M T W T F S

Foreman:

Project Details

Weather

High	Low

Weather Conditions

Ground Conditions

Safety Meeting

Yes	No	Subject:

Concerns:

Visitors

Schedule

Completion Date	
Days Ahead	
Days Behind	

Accidents / Injuries / Delays

Summary of Work Done Today

Signature: Title:

Employee	Trade	Hours	OT	Subcontractors	Trade	OT

Equipment	#	On	Off

Material Delivered	#	Rentals	Rented From/Rate

DATE:		DAY:	S	M	T	W	T	F	S

Foreman:

Project Details	Weather

	High	Low

Safety Meeting

Yes	No	Subject:

Concerns:

Weather Conditions

Ground Conditions

Visitors

Schedule

Completion Date	
Days Ahead	
Days Behind	

Accidents / Injuries / Delays

Summary of Work Done Today

Signature:		Title:

Employee	Trade	Hours	OT	Subcontractors	Trade	OT

Equipment	#	On	Off

Material Delivered	#	Rentals	Rented From/Rate

DATE:	DAY:	S	M	T	W	T	F	S

Foreman:

Project Details

Weather

High	Low

Weather Conditions

Ground Conditions

Safety Meeting

Yes	No	Subject:

Concerns:

Visitors

Schedule

Completion Date	
Days Ahead	
Days Behind	

Accidents / Injuries / Delays

Summary of Work Done Today

Signature: Title:

Employee	Trade	Hours	OT	Subcontractors	Trade	OT

Equipment	#	On	Off

Material Delivered	#	Rentals	Rented From/Rate

DATE:		DAY:	S M T W T F S

Foreman:

Project Details

Weather

High	Low

Weather Conditions

Safety Meeting

Yes	No	Subject:

Concerns:

Ground Conditions

Visitors

Schedule

Completion Date	
Days Ahead	
Days Behind	

Accidents / Injuries / Delays

Summary of Work Done Today

Signature:	Title:

Employee	Trade	Hours	OT	Subcontractors	Trade	OT

Equipment	#	On	Off

Material Delivered	#	Rentals	Rented From/Rate

DATE:		DAY:	S	M	T	W	T	F	S

Foreman:

Project Details	Weather	
	High	Low
	Weather Conditions	

Safety Meeting

Yes	No	Subject:
		Concerns:

Ground Conditions

Visitors

Schedule

Completion Date	
Days Ahead	
Days Behind	

Accidents / Injuries / Delays

Summary of Work Done Today

Signature: Title:

Employee	Trade	Hours	OT	Subcontractors	Trade	OT

Equipment	#	On	Off

Material Delivered	#	Rentals	Rented From/Rate

DATE:		DAY:	S M T W T F S

Foreman:

Project Details

Weather

High	Low

Weather Conditions

Safety Meeting

Yes	No	Subject:

Concerns:

Ground Conditions

Visitors

Schedule

Completion Date	
Days Ahead	
Days Behind	

Accidents / Injuries / Delays

Summary of Work Done Today

Signature:	Title:

Employee	Trade	Hours	OT	Subcontractors	Trade	OT

Equipment	#	On	Off

Material Delivered	#	Rentals	Rented From/Rate

DATE:		DAY:	S	M	T	W	T	F	S

Foreman:

Project Details

Weather

High	Low

Weather Conditions

Ground Conditions

Safety Meeting

Yes	No	Subject:

Concerns:

Visitors

Schedule

Completion Date	
Days Ahead	
Days Behind	

Accidents / Injuries / Delays

Summary of Work Done Today

Signature: Title:

Employee	Trade	Hours	OT	Subcontractors	Trade	OT

Equipment	#	On	Off

Material Delivered	#	Rentals	Rented From/Rate

DATE:		DAY:	S	M	T	W	T	F	S

Foreman:

Project Details	Weather	
	High	Low

Safety Meeting	Weather Conditions
Yes	No	Subject:	

Concerns:

	Ground Conditions

Visitors	Schedule	
	Completion Date	
	Days Ahead	
	Days Behind	

Accidents / Injuries / Delays

Summary of Work Done Today

Signature: Title:

Employee	Trade	Hours	OT	Subcontractors	Trade	OT

Equipment	#	On	Off

Material Delivered	#	Rentals	Rented From/Rate

DATE:		DAY:	S	M	T	W	T	F	S

Foreman:

Project Details	Weather

		High	Low

Weather Conditions

Safety Meeting

Yes	No	Subject:

Concerns:

Ground Conditions

Visitors

Schedule

Completion Date	
Days Ahead	
Days Behind	

Accidents / Injuries / Delays

Summary of Work Done Today

Signature: Title:

Employee	Trade	Hours	OT	Subcontractors	Trade	OT

Equipment	#	On	Off

Material Delivered	#	Rentals	Rented From/Rate

DATE:		DAY:	S M T W T F S

Foreman:

Project Details	Weather	
	High	Low

Weather Conditions

Ground Conditions

Safety Meeting

Yes	No	Subject:

Concerns:

Visitors

Schedule

Completion Date	
Days Ahead	
Days Behind	

Accidents / Injuries / Delays

Summary of Work Done Today

Signature:	Title:

Employee	Trade	Hours	OT	Subcontractors	Trade	OT

Equipment	#	On	Off

Material Delivered	#	Rentals	Rented From/Rate

DATE:		DAY:	S	M	T	W	T	F	S

Foreman:

Project Details

	Weather	
	High	Low

Weather Conditions

Ground Conditions

Safety Meeting

Yes	No	Subject:

Concerns:

Visitors

Schedule

Completion Date	
Days Ahead	
Days Behind	

Accidents / Injuries / Delays

Summary of Work Done Today

Signature: Title:

Employee	Trade	Hours	OT	Subcontractors	Trade	OT

Equipment	#	On	Off

Material Delivered	#	Rentals	Rented From/Rate

DATE:	DAY:	S M T W T F S

Foreman:

Project Details

Weather

High	Low

Weather Conditions

Ground Conditions

Safety Meeting

Yes	No	Subject:

Concerns:

Visitors

Schedule

Completion Date	
Days Ahead	
Days Behind	

Accidents / Injuries / Delays

Summary of Work Done Today

Signature:	Title:

Employee	Trade	Hours	OT	Subcontractors	Trade	OT

Equipment	#	On	Off

Material Delivered	#	Rentals	Rented From/Rate

DATE:		DAY:	S M T W T F S

Foreman:

Project Details	Weather

		High	Low

Weather Conditions

Ground Conditions

Safety Meeting

Yes	No	Subject:
Concerns:		

Visitors	Schedule

Completion Date	
Days Ahead	
Days Behind	

Accidents / Injuries / Delays

Summary of Work Done Today

Signature: Title:

Employee	Trade	Hours	OT	Subcontractors	Trade	OT

Equipment	#	On	Off

Material Delivered	#	Rentals	Rented From/Rate

DATE:		DAY:	S M T W T F S

Foreman:

Project Details	Weather	
	High	**Low**

	Weather Conditions
Safety Meeting	

Yes	No	Subject:
Concerns:		

	Ground Conditions

Visitors	Schedule	
	Completion Date	
	Days Ahead	
	Days Behind	

Accidents / Injuries / Delays

Summary of Work Done Today

Signature: Title:

Employee	Trade	Hours	OT	Subcontractors	Trade	OT

Equipment	#	On	Off

Material Delivered	#	Rentals	Rented From/Rate

DATE:		DAY:	S	M	T	W	T	F	S

Foreman:

Project Details

Weather

High	Low

Weather Conditions

Ground Conditions

Safety Meeting

Yes	No	Subject:

Concerns:

Visitors

Schedule

Completion Date	
Days Ahead	
Days Behind	

Accidents / Injuries / Delays

Summary of Work Done Today

Signature:	Title:

Employee	Trade	Hours	OT	Subcontractors	Trade	OT

Equipment	#	On	Off

Material Delivered	#	Rentals	Rented From/Rate

DATE:		DAY:	S M T W T F S

Foreman:

Project Details

Weather

High	Low

Weather Conditions

Ground Conditions

Safety Meeting

Yes	No	Subject:

Concerns:

Visitors

Schedule

Completion Date	
Days Ahead	
Days Behind	

Accidents / Injuries / Delays

Summary of Work Done Today

Signature: Title:

Employee	Trade	Hours	OT	Subcontractors	Trade	OT

Equipment	#	On	Off

Material Delivered	#	Rentals	Rented From/Rate

DATE: **DAY:**   S   M   T   W   T   F   S

Foreman:

Project Details

Weather

High	Low

Weather Conditions

Ground Conditions

Safety Meeting

Yes	No	Subject:

Concerns:

Visitors

Schedule

Completion Date	
Days Ahead	
Days Behind	

Accidents / Injuries / Delays

Summary of Work Done Today

Signature: Title:

Employee	Trade	Hours	OT	Subcontractors	Trade	OT

Equipment	#	On	Off

Material Delivered	#	Rentals	Rented From/Rate

| DATE: | | DAY: | S M T W T F S |

Foreman:

Project Details

Weather

High	Low

Weather Conditions

Ground Conditions

Safety Meeting

Yes	No	Subject:

Concerns:

Visitors

Schedule

Completion Date	
Days Ahead	
Days Behind	

Accidents / Injuries / Delays

Summary of Work Done Today

Signature: Title:

Employee	Trade	Hours	OT	Subcontractors	Trade	OT

Equipment	#	On	Off

Material Delivered	#	Rentals	Rented From/Rate

DATE:		DAY:	S	M	T	W	T	F	S

Foreman:

Project Details	Weather

	High	Low

Weather Conditions

Safety Meeting

Yes	No	Subject:

Concerns:

Ground Conditions

Visitors

Schedule

Completion Date	
Days Ahead	
Days Behind	

Accidents / Injuries / Delays

Summary of Work Done Today

Signature: Title:

Employee	Trade	Hours	OT	Subcontractors	Trade	OT

Equipment	#	On	Off

Material Delivered	#	Rentals	Rented From/Rate

DATE:		DAY:	S	M	T	W	T	F	S

Foreman:

Project Details	Weather

		High	Low

Weather Conditions

Safety Meeting

Yes	No	Subject:

Concerns:

Ground Conditions

Visitors

Schedule

Completion Date	
Days Ahead	
Days Behind	

Accidents / Injuries / Delays

Summary of Work Done Today

Signature: Title:

Employee	Trade	Hours	OT	Subcontractors	Trade	OT

Equipment	#	On	Off

Material Delivered	#	Rentals	Rented From/Rate

DATE:	DAY:	S	M	T	W	T	F	S

Foreman:

Project Details

Weather

High	Low

Weather Conditions

Ground Conditions

Safety Meeting

Yes	No	Subject:

Concerns:

Visitors

Schedule

Completion Date	
Days Ahead	
Days Behind	

Accidents / Injuries / Delays

Summary of Work Done Today

Signature: Title:

Employee	Trade	Hours	OT	Subcontractors	Trade	OT

Equipment	#	On	Off

Material Delivered	#	Rentals	Rented From/Rate

DATE:	DAY:	S	M	T	W	T	F	S

Foreman:

Project Details	Weather	
	High	Low

	Weather Conditions

Safety Meeting

Yes	No	Subject:
Concerns:		

Ground Conditions

Visitors

Schedule

Completion Date	
Days Ahead	
Days Behind	

Accidents / Injuries / Delays

Summary of Work Done Today

Signature:	Title:

Employee	Trade	Hours	OT	Subcontractors	Trade	OT

Equipment	#	On	Off

Material Delivered	#	Rentals	Rented From/Rate

DATE:		DAY:	S	M	T	W	T	F	S

Foreman:

Project Details

Weather

High	Low

Weather Conditions

Ground Conditions

Safety Meeting

Yes	No	Subject:

Concerns:

Visitors

Schedule

Completion Date	
Days Ahead	
Days Behind	

Accidents / Injuries / Delays

Summary of Work Done Today

Signature:	Title:

Employee	Trade	Hours	OT	Subcontractors	Trade	OT

Equipment	#	On	Off

Material Delivered	#	Rentals	Rented From/Rate

DATE:		DAY:	S	M	T	W	T	F	S

Foreman:

Project Details	Weather

	Weather	
	High	Low

Weather Conditions

Safety Meeting

Yes	No	Subject:
Concerns:		

Ground Conditions

Visitors

Schedule

Completion Date	
Days Ahead	
Days Behind	

Accidents / Injuries / Delays

Summary of Work Done Today

Signature: Title:

Employee	Trade	Hours	OT	Subcontractors	Trade	OT

Equipment	#	On	Off

Material Delivered	#	Rentals	Rented From/Rate

DATE:		DAY:	S	M	T	W	T	F	S

Foreman:

Project Details

Weather

High	Low

Weather Conditions

Ground Conditions

Safety Meeting

Yes	No	Subject:

Concerns:

Visitors

Schedule

Completion Date	
Days Ahead	
Days Behind	

Accidents / Injuries / Delays

Summary of Work Done Today

Signature: Title:

Employee	Trade	Hours	OT	Subcontractors	Trade	OT

Equipment	#	On	Off

Material Delivered	#	Rentals	Rented From/Rate